YOUR KNOWLEDGE HAS VALUE

- We will publish your bachelor's and master's thesis, essays and papers

- Your own eBook and book - sold worldwide in all relevant shops

- Earn money with each sale

Upload your text at www.GRIN.com and publish for free

Wasihun Regu

Allelopathic Effect of "Eucalyptus camaldulensis" and "Eucalyptus grandis" on "Phaseolus vulgaris" and "Zea mays"

Tree Crop Interaction

GRIN Verlag

Bibliografische Information der Deutschen Nationalbibliothek:

Die Deutsche Bibliothek verzeichnet diese Publikation in der Deutschen National-
bibliografie; detaillierte bibliografische Daten sind im Internet über http://dnb.d-
nb.de/ abrufbar.

Imprint:

Copyright © 2012 GRIN Verlag GmbH
Druck und Bindung: Books on Demand GmbH, Norderstedt Germany
ISBN: 978-3-656-69992-7

This book at GRIN:

http://www.grin.com/en/e-book/276488/allelopathic-effect-of-eucalyptus-camaldu-
lensis-and-eucalyptus-grandis

GRIN - Your knowledge has value

Der GRIN Verlag publiziert seit 1998 wissenschaftliche Arbeiten von Studenten, Hochschullehrern und anderen Akademikern als eBook und gedrucktes Buch. Die Verlagswebsite www.grin.com ist die ideale Plattform zur Veröffentlichung von Hausarbeiten, Abschlussarbeiten, wissenschaftlichen Aufsätzen, Dissertationen und Fachbüchern.

Visit us on the internet:

http://www.grin.com/

http://www.facebook.com/grincom

http://www.twitter.com/grin_com

Allelopathic Effect of *Eucalyptus camaldulensis* and *Eucalyptus grandis* on *Phaseolus vulgaris* and *Zea mays*

*Wasihun Regu Gurmu
*BSc in Natural Resource Management and MSc in Agroforestry
Areka Agricultural Research Center, Areka, Ethiopia. P.O.Box: 79.

ABSTRACT

Allelopathic effects of Eucalyptus are widely reported and are considered the major factor limiting the cultivation of agricultural crops intercropped with Eucalyptus stands. However, reports of the allelopathic effects of eucalyptus are mostly based on laboratory bioassay, not on field trials. The objective of the present study was to determine the effects of different leaf powder proportions of Eucalyptus species on agricultural crops as well as change in soil reaction. This research was undertaken in a pot experiment grown in the nursery of Hawassa College of Agriculture campus laid in a completely randomized design containing leaf powder prepared by drying and grinding fresh collected leaves of Eucalyptus camaldulensis and Eucalyptus grandis and containing two agricultural crops (haricot bean and maize). Results suggested that leaf powder of each tree species induced significant inhibitory effect on emergence and growth of each crops compared to the control treatments. The study revealed that application of low-dose leaf powder of either tree species had low effect on all measured parameters of both crops. Eucalyptus grandis has more inhibitory effect on both crops than Eucalyptus camaldulensis, and the negative effect of allelochemicals is more pronounced on haricot bean than maize in both tree species. Thus, it is recommended that maize and haricot bean should not be planted very close or not to Eucalyptus trees. In addition, crops seed rate should be increased to get maximum emergence.

key words: Allelopathy, *Eucalyptus camaldulensis*, *Eucalyptus grandis*, Emergence, Seedling growth, *Zea mays*, *Phaseolus vulgaris*

INTRODUCTION

Forest plantations in Ethiopia are mainly monocultures of exotic species. Nevertheless, they are predominantly monocultures of *Eucalyptus species* which are estimated to cover about 93% of the total plantation area in the country (FAO, 1981).

Nowadays many people in Ethiopia are dependent on *Eucalyptus* for fuel wood, construction wood and income generation (Davidson, 1995). The sale of *Eucalyptus* poles and products has substantial potential to raise farm incomes, reduce poverty, increase food security, and diversify smallholder-farming systems in less-favored areas (Jagger and Pender, 2000). The other greatest positive contribution of *Eucalyptus* is perhaps in replacing indigenous species for firewood, thereby, preventing further denudation of natural forests (Evans, 1992). Generally, *Eucalyptus* in Ethiopia is

1

highly favored for planting near and outside the croplands and it is common to see the species in most of the agroecological zones of the country.

While *Eucalyptus* have many benefits, of all widely used plantation species they have attracted by far the most criticism (Evans, 1992). Some of the critics associated with it are: the species do not provide organic matter and depletes soil nutrients needed by agricultural crops, it depletes water resources and competes with agricultural crops, it suppresses ground vegetation and resulting unsuitability to soil erosion control, the leaves of *Eucalyptus* are not palatable and cannot be used as fodder species, and it is believed to have allelopathic effect (Jagger and Pender, 2000).

Eucalyptus species effect on the environment, on undergrowth vegetation and soil fertility may vary within different geographical areas, rainfall regimes and between species. So to discourage or promote the planting of *Eucalyptus* and to use them for agroforestry purposes, sufficient scientific evidences on the ecological impact, on undergrowth vegetation, soil fertility, and the quality of the product should be further investigated (Kumar *et al.*, 2008).

Decline in crop yields in cropping and agroforestry system in recent years has been attributed to allelopathic effects (Oyun, 2006). In the traditional agroforestry system of southern Ethiopia, people are growing several tree species in or around the agricultural fields as a boundary or woodlot. Among these are *E. grandis* and *E. camaldulensis* and due to their adverse effects some of the farmers are now reluctant to grow these tree species in their agricultural fields (CNST, 1997).

Today, Haricot bean *(Phaseolus vulgaris* L.) and Maize (*Zea mays* L.) are grown on more land and continues to be the most important food grain source for humans. Haricot bean and Maize are one of the top priority crops and the basic staple food crops of the majority of the population in the southern region of Ethiopia. So the purpose of the present study was to elucidate the allelopathic potential of different leaf powder proportions of *Eucalyptus camaldulensis* and *Eucalyptus grandis* on haricot bean and maize crops.

MATERIALS AND METHODS

The study was conducted in the nursery site of Hawassa College of Agriculture, Hawassa University, Ethiopia. *Eucalyptus* species were considered from the plantation at WondoGenet College of Forestry and Natural Resources, Ethiopia.

Randomly collecting fresh leaves of 2 to 3-year old *Eucalyptus grandis* and *Eucalyptus camaldulensis* from the top, middle and bottom canopy and they were dried in an open air under shade, and after two days, the leaves were then put in oven with a temperature of 50 ^{0}C for 48 hours (El-Khawas and Shehata, 2005). All the leaf samples were ground with electric grinder. The prepared powder was stored in plastic bags at room temperature until its application into pots. Seeds of maize (ACV6 variety) and haricot bean (Melkasa1 variety) were obtained from Hawassa College of Agriculture, Hawassa University and used in the study as recipient crops.

2

Experimental design and layout

The experiment was laid down in a Complete Randomized Design (CRD). *Eucalyptus camaldulensis* and *Eucalyptus grandis* were considered as the donor plants and the receptor or companion agricultural crops considered were haricot bean and maize.

Pots were prepared and arranged in rows independently for both crops and fairly leveled by adding local top soil collected from 0-30 cm depth. All weeds and other debris were removed and treated with 0, 10, 50, 100, 150 and 200 g/pot of *Eucalyptus* leaf powder (Ahmed *et al.*, 2008). All the treatments were replicated four times consisting a total of 96 pots/experimental units.

Equal amount of water was added into all the plastic pots and kept for one week in the net house provided with plastic sheets to develop any possible microbial activity prior to sowing. The seeds of both crops were floated in water to eliminate the empty ones, and then four seeds of haricot bean and four grains of maize were planted in each of the treated plastic pots at 42 cm depth, 30 cm top and 20 cm bottom diameters provided with drainage holes. The pots were watered every two days on the average with normal tap water. After emergence, seedlings were thinned to two plants per pot. The experiment continued over a period of 21 days in which the final measurements were recorded.

The grounded leaves weighing a total of 4.08 kg of each of the two *Eucalyptus* species were mixed with pot soil uniformly in the following treatment proportions having a separate experiments for each crops as:

Set I: Treatments to assess allelopathic effect of *Eucalyptus grandis* on haricot bean and maize

T_0= haricot bean + 0 g/pot (Control) T_0= maize + 0 g/pot (Control)

T_1= haricot bean + 10 g/pot T_1= maize + 10 g/pot

T_2= haricot bean + 50 g/pot T_2= maize + 50 g/pot

T_3= haricot bean + 100 g/pot T_3= maize + 100 g/pot

T_4= haricot bean + 150 g/pot T_4= maize + 150 g/pot

T_5= haricot bean + 200 g/pot T_5= maize + 200 g/pot

Set II: Treatments to assess allelopathic effect of *Eucalyptus camaldulensis* on haricot bean and maize

T_0= haricot bean + 0 g/pot (Control) T_0= maize + 0 g/pot (Control)

T_1= haricot bean + 10 g/pot T_1= maize + 10 g/pot

T_2= haricot bean + 50 g/pot T_2= maize + 50 g/pot;

T_3= haricot bean + 100 g/pot T_3= maize + 100 g/pot;

T_4= haricot bean + 150 g/pot T_4= maize + 150 g/pot;

T_5= haricot bean + 200 g/pot T_5= maize + 200 g/pot;

Data Collection and Statistical Analysis

Seed emergence (%), Shoot length (cm), Root length (cm), Shoot fresh and Shoot dry weight (g), Root fresh and Root dry weight (g), Collar diameter (cm), and pH of soil were measred following previous publications by (Rho and Kil, 1986, Bewley and Black, 1994, Saxenaet al., 1996, El-Darier, 2002 and Khan, 2003).

The emergence percentage and percent inhibition or stimulation was calculated as:

- Emergence (%) = (No of emerged seeds / Total No of seeds sown) × 100

- Percent inhibition or Stimulation = 100 - (Treatment value/Control value) × 100

For shoot length (cm) two plants in each treatment were measured with meter rod and the measurements were taken from the stem joint to the tip of the terminal leaf.

For root length (cm) two plants in each treatment were measured with meter rod and the measurements were taken from the root joint to the tip of the tap/main root. All the measurements were summed up and divided by the number of plants measured to find out the single plant average shoot and root length. Collar diameters (cm) of both crops were determined using a caliper.

After the different organs were separated to shoot and root biomass, they were kept in paper bags and tagged properly. Wet and dry weights of the biomass were measured using scientific electrical balance.

For shoot and root fresh weight (g) two fresh plant shoots and roots were weighed and the total weight was averaged to determine single plant shoot and root fresh weight.

For shoot and root dry weight (g) the two fresh plants shoots and roots were dried in oven at 65 °c for 24 hours then weighed, and the total was averaged to determine single plant shoot and root dry weight.

After both crops were harvested soil samples were prepared from each pot; the samples that weighed 10 g each were dried and sieved through a 2 mm mesh sieve and added into 100 ml beakers. 25 ml distilled water was added from a measuring cylinder for 1:2.5 soil/water suspension. Stirring was done for 1 minute by using a glass stirrer and allow the sample to equilibrate and pH were measured with pH meter after 1 hour on the upper part of the suspension.

The analysis of variance (ANOVA) was conducted using the general linear model (PROC GLM) procedure in Statistical Analysis System (SAS) program version 9. Means were compared for all significant parameters using Least Significant Difference (LSD) test at the 5 % level of significance (Mkula, 2006).

4

RESULTS

Effect of *Eucalyptus grandis* on haricot bean and maize

The results of seed emergence showed that majority of treatment levels of *Eucalyptus grandis* significantly ($p \leq 0.05$) reduced maize and haricot bean seed emergence (Table 1 and 2) except treatment levels of 10 and 50 g/pot in maize the rest showed significant effect compared to the control treatment, and a 200 g/pot showed statistical similar with 150 g/pot. For haricot bean all treatments are statistically significant compared to the control and there is no significant different between all treatment levels.

The results of shoot length showed that all treatment levels of *Eucalyptus grandis* significantly ($p \leq 0.05$) reduced both maize and haricot bean shoot length as compared to the control treatment (Table 1 and 2). For haricot bean all treatments are statistically significant compared to the control and there is no significant different between 10, 50 and 100 g/pot treatment levels. A 200 g/pot and 150 g/pot showed statistically similar and significantly different from the rest treatments. For maize all treatments are statistically significant compared to the control and there is no significant different between 10 and 50 g/pot treatment levels. The smallest mean shoot length was observed in the 200 g/pot treated pots for both crops.

The results of crop root length showed that all treatment levels of *Eucalyptus grandis* significantly ($p \leq 0.05$) reduced both maize and haricot bean root length except for the 10 g/pot treatment in maize as compared to the control treatment (Table 1 and 2). The smallest mean root lengths were observed in the 200 g/pot and 150 g/pot treated pots for haricot bean and maize respectively.

5

Table 1: Response to allelopathic effect of *Eucalyptus grandis* on emergence and growth of haricot bean seedlings

Treatments	Emergence (%)	Shoot length (cm)	Root length (cm)	Collar diameter (cm)	Shoot fresh weight (g)	Root fresh weight (g)	Shoot dry weight (g)	Root dry weight (g)	Soil reaction (pH)
Control	91.67a	16.10a	17.00a	0.41a	3.61a	0.19a	1.00a	0.13a	8.08a
10g/pot	41.67b (-54.5)	9.98b (-38)	10.50b (-38.2)	0.35a (-14.5)	1.25b (-65.4)	0.10b (-48.1)	0.46b (-54.5)	0.07b (-48.8)	7.75b (-4)
50g/pot	33.33b (-63.6)	8.50b (-47.2)	9.25bc (-45.6)	0.34a (-18.8)	0.96b (-73.5)	0.07bc (-66.2)	0.29bc (-71.5)	0.05bc (-65)	7.63bc (-5.6)
100g/pot	33.33b (-63.6)	6.38bc (-60.4)	6.88bc (-59.6)	0.20b (-51.5)	0.38b (-89.5)	0.03cd (-82.3)	0.23bc (-77.3)	0.01c (-91.3)	7.53bc (-6.8)
150g/pot	33.33b (-63.6)	4.25cd (-73.6)	5.88bc (-65.4)	0.13bc (-68.8)	0.19b (-94.7)	0.02cd (-88.8)	0.16bc (-84.5)	0.01c (-94.4)	7.45bc (-7.7)
200g/pot	33.33b (-63.6)	2.15d (-86.6)	3.75c (-77.9)	0.06c (-85.2)	0.11b (-97)	0.012d (-95.3)	0.09c (-90.1)	0.01c (-95)	7.38c (-8.7)
LSD @ p≤0.05	14.29	4.17	6.43	0.12	1.82	0.05	0.33	0.05	0.30

*Means in the same column followed by the same letter are not significantly different according to LSD at probability level of 0.05. Value in the parenthesis indicates % inhibitory (-) effects in comparison to control.

Table 2: Response to allelopathic effect of *Eucalyptus grandis* one emergence and growth of maize seedlings

Treatments	Emergence (%)	Shoot length (cm)	Root length (cm)	Collar diameter (cm)	Shoot fresh weight (g)	Root fresh weight (g)	Shoot dry weight (g)	Root dry weight (g)	Soil reaction (pH)
Control	100a	26.13a	46.25a	0.91a	10.9a	2.45a	2.55a	1.88a	8.03a
10g/pot	100a	20.25b (-22.5)	38.75a (-16.2)	0.75a (-17.8)	7.34ab (-32.8)	1.09b (-55.3)	1.55b (-39.2)	0.72b (-61.5)	7.85ab (-2.2)
50g/pot	91.67a (-8.3)	15.88b (-39.2)	25.50b (-44.9)	0.53b (-41.9)	4.83bc (-55.8)	0.68bc (-72.2)	0.88c (-65.7)	0.38c (-79.6)	7.80ab (-2.8)
100g/pot	66.67b (-33.3)	10.35c (-60.4)	16.88bc (-63.5)	0.42bc (-54.5)	1.42c (-87.1)	0.17cd (-93.1)	0.18d (-93.1)	0.13cd (-93.2)	7.63b (-4.9)
150g/pot	50.00bc (-50)	6.75cd (-74.2)	10.38c (-77.6)	0.33bc (-64.1)	0.71c (-93.5)	0.13d (-94.7)	0.11d (-95.7)	0.06d (-96.9)	7.50b (-6.5)
200g/pot	33.33c (-66.7)	5.20d (-80.1)	12.38c (-73.2)	0.21c (-76.9)	0.52c (-95.2)	0.07d (-97.1)	0.04d (-98.3)	0.04d (-97.9)	7.10c (-11.5)
LSD @ p≤0.05	22.60	5.08	12.64	0.22	4.38	0.51	0.66	0.27	0.39

*Means in the same column followed by the same letter are not significantly different according to LSD at probability level of 0.05. Value in the parenthesis indicates % inhibitory (-) effects in comparison to control.

The results on collar diameter showed that majority of treatment levels of *Eucalyptus grandis* significantly ($p \leq 0.05$) reduced both maize and haricot bean collar diameter except for the 10, and 50 g/pot levels compared to the control treatment (Table 1 and 2). The smallest mean collar diameter was observed in the 200 g/pot treated pots for both crops.

The results showed that all treatment levels of *Eucalyptus grandis* significantly ($p \leq 0.05$) reduced both maize and haricot bean shoot, root fresh and shoot, root dry weight except for the 10 g/pot in maize shoot fresh weight as compared to the control treatment (Table 1 and 2). The smallest mean shoot, root fresh and shoot, root dry weight was observed in the 200 g/pot treated pots for both crops.

The results of soil pH showed that all treatment levels of *Eucalyptus grandis* significantly ($p \leq 0.05$) reduced soil pH in maize and haricot bean pot except for the 10, and 50 g/pot in the maize pot as compared to the control treatment (Table 1 and 2). The smallest mean soil pH was observed in the 200 g/pot treated pots for both crops.

Effect of *Eucalyptus camaldulensis* on haricot bean and maize

The results of seed emergence showed that all treatment levels of *Eucalyptus camaldulensis* significantly ($p \leq 0.05$) reduced maize and haricot bean seed emergence except for the 10 g/pot in haricot bean as compared to the control treatment (Table 3 and 4). The smallest mean seed emergence and statistically similar with 150 g/pot was observed in the 200 g/pot treated pots for maize. The smallest mean seed emergence was observed in the 150 and 200 g/pot treated pots for haricot bean and in the 200 g/pot treated pots for maize.

The result of shoot length showed that all treatment levels of *Eucalyptus camaldulensis* significantly ($p \leq 0.05$) reduced both maize and haricot bean shoot length except for the 10 g/pot in haricot bean as compared to the control treatment (Table 3 and 4). Even if there is a statistical similarity between 100, 150 and 200 g/pot treatments for haricot bean, and between 150 and 200 g/pot treatments for maize the smallest mean shoot length was observed in the 200 g/pot treated pots for both crops.

The results of root length showed that all treatment levels of *Eucalyptus camaldulensis* significantly ($p \leq 0.05$) reduced both maize and haricot bean root length as compared to the control treatment (Table 3 and 4). The smallest mean root length was observed in the 200 g/pot treated pots for both crops.

Table 3: Response to allelopathic effect of *Eucalyptus camaldulensis* on emergence and growth of haricot bean seedlings

Treatments	Emergence (%)	Shoot length (cm)	Root length (cm)	Collar diameter (cm)	Shoot fresh weight (g)	Root fresh weight (g)	Shoot dry weight (g)	Root dry weight (g)	Soil reaction (pH)
Control	100a	16.50a	22a	0.43a	3.69a	0.27a	1.08a	0.17a	8.08a
10g/pot	91.67ab (-8.3)	14.68ab (-11.1)	14.38b (-34.6)	0.38ab (-10.6)	2.08ab (-44.9)	0.13b (-53.3)	0.75ab (-30.2)	0.08b (-55.1)	7.93ab (-1.9)
50g/pot	75.00b (-25)	12.43bc (-24.7)	9.43bc (-57.2)	0.32bc (-24.1)	1.45b (-60.9)	0.11bc (-59.9)	0.47bc (-56.5)	0.04b (-74.9)	7.85b (-2.8)
100g/pot	41.67c (-58.3)	10.48cd (-36.5)	7.98bc (-63.8)	0.28cd (-35.3)	1.16b (-68.7)	0.08bc (-71.9)	0.37bc (-66)	0.03b (-85.5)	7.75bc (-4)
150g/pot	33.33c (-66.7)	10.13cd (-38.6)	7.38bc (-66.5)	0.21d (-50)	0.86b (-76.7)	0.04bc (-84.9)	0.28c (-73.7)	0.02b (-88.1)	7.63cd (-5.6)
200g/pot	33.33c (-66.7)	7.13d (-56.8)	5.13c (-76.7)	0.18d (-57.1)	0.55b (-85.1)	0.02c (-93.4)	0.16c (-85.3)	0.01b (-94.2)	7.45d (-7.7)
LSD @ p≤0.05	24.06	4.06	7.41	0.09	1.69	0.09	0.42	0.07	0.19

*Means in the same column followed by the same letter are not significantly different according to LSD at probability level of 0.05. Value in the parenthesis indicates % inhibitory (-) effects in comparison to control.

Table 4: Response to allelopathic effect of *Eucalyptus camaldulensis* on emergence and growth of maize seedlings

Treatments	Emergence (%)	Shoot length (cm)	Root length (cm)	Collar diameter (cm)	Shoot fresh weight (g)	Root fresh weight (g)	Shoot dry weight (g)	Root dry weight (g)	Soil reaction (pH)
Control	91.67a	30.25a	50.75a	1.04a	18.62a	2.45a	3.30a	1.88a	7.85a
10g/pot	91.67a	25.25b (-16.5)	40.38b (-20.4)	0.91a (-13.2)	11.49b (-38.3)	1.88b (-23.1)	2.38b (-28)	0.98b (-48)	7.70a (-1.9)
50g/pot	83.33a (-9.1)	20.63c (-31.8)	40.50b (-20.2)	0.75b (-28.3)	7.89bc (-57.6)	1.02c (-58.4)	1.63bc (-50.8)	0.65c (-65.3)	7.50ab (-4.5)
100g/pot	83.33a (-9.1)	15.63d (-48.3)	29.25c (-42.3)	0.64b (-38.8)	4.62cd (-75.2)	0.69cd (-71.7)	0.98cd (-70.5)	0.45cd (-76)	7.45ab (-5.1)
150g/pot	66.67ab (-27.3)	9.98e (-67)	16.33d (-67.8)	0.46c (-56.4)	1.45d (-92.2)	0.25e (-89.8)	0.33d (-90.2)	0.22de (-88.3)	7.40ab (-5.7)
200g/pot	41.67b (-54.5)	7.25e (-76)	12 d (-76.4)	0.35c (-66.7)	0.78d (-95.8)	0.22de (-91.2)	0.13d (-96.2)	0.07e (-96.1)	7.23b (-7.9)
LSD @ p≤0.05	29.18	4.39	8.08	0.14	4.57	0.46	0.86	0.29	0.45

*Means in the same column followed by the same letter are not significantly different according to LSD at probability level of 0.05. Value in the parenthesis indicates % inhibitory (-) effects in comparison to control

The results on collar diameter showed that the majority of treatment levels of *Eucalyptus camaldulensis* significantly ($p \leq 0.05$) reduced both maize and haricot bean collar diameter except for 10 g/pot in both crops as compared to the control treatment (Table 3 and 4). The smallest mean collar diameter was observed in the 200 g/pot treated pots for both crops, even though it has a statistical similarity with 100 and 150 g/pot in haricot bean and with 150 g/pot in maize.

The results showed that all treatment levels of *Eucalyptus camaldulensis* significantly ($p \leq 0.05$) reduced both maize and haricot bean shoot, root fresh and shoot, root dry weight except for the 10g/pot in haricot bean shoot fresh weight and shoot dry weight as compared to the control treatment (Table 3 and 4). The smallest mean shoot, root fresh and shoot, root dry weight was observed in the 200 g/pot treated pots for both crops, even though it has statistical similarity with all treatments in haricot bean shoot fresh and root dry weight and except with a 10 g/pot of root fresh and shoot dry weight.

The results of soil pH showed that except treatment 10 g/pot, the rest all treatment levels of *Eucalyptus camaldulensis* significantly ($p \leq 0.05$) reduced soil pH in haricot bean and maize pots as compared to the control treatments (Table 3 and 4). More over all treatments were statistical similar except for treatment level of 200 g/pot where significantly the smallest soil pH was observed for both crops.

DISCUSSION

The present study revealed that no treatments of both *Eucalyptus* species had stimulating effect on all measured parameters of either test crops except no noticeable effect in the treatment of 10 g/pot of *Eucalyptus grandis* on maize emergence as compared to control. Probable reasons for this could be the elevated amount of allelochemicals and inhibitory effect present in each *Eucalyptus* species leaf powder that affected the essential growth processes. From the mean results, the effect is increased with the increase of leaf powder application. Similar studies elsewhere have found that inhibitory effect of *Eucalyptus camaldulensis* depended on concentration of extract and litter-fall with higher concentration of the materials, having higher effects and vice versa (Ahmed*et al.*, 2008). These results are also in agreement with those reported by (Lisanework and Michelson, 1993) who noted a decrease in germination of maize due to applied *Eucalyptus globules, Eucalyptus camaldulensis* and *Eucalyptus saliga* extract. The results validate the findings of (Ebrahim*et al.,*1999, Khan *et al.,*1999 and Khan *et al.*, 2007) who reported that leaf extract of *Eucalyptus camaldulensis* and *Eucalyptus microthecia* delayed and inhibited germination of maize. Thakur and Bhardwaj (1992) also reported that leachates from *E.globulus* leave significantly reduce maize germination. Rao and Reddy (1984) reported inhibition of germination and other growth parameters in horse gram, green gram, cowpeas and beans due to the leaf extracts of *Eucalyptus camaldulensis*. Shivanna*et al* (1992) also reported the allelopathic effect of *Eucalyptus camaldulensis* on rag (*Eleusinecoracana*), Cowpeas (*Vignaungnicalanta*) and Sesmum (*Sesmumindicum*).

The highest reduction in the mean shoot length of haricot bean and maize was observed in the 200 g/pot of *Eucalyptus grandis* treated pots with values of 86.6% and 80.1%. Similarly the highest reductions in the mean shoot length were observed in the 200 g/pot of *Eucalyptus camaldulensis* treated pots with 56.8% and 76% for haricot bean and maize respectively. Schumann *et al* (1995) also reported that water

11

extracts of *E. grandis* significantly reduced weed establishment. Swaminathan*et al* (1993) evaluated the allelopathic effect of eight multipurpose trees including *Eucalyptus tereticornis* on maize, red gram and sesame. All the trees inhibited germination and growth of all the crops. Khan *et al* (1999) also reported that *Eucalyptus camaldulensis* extracts reduced maize seedling height and fresh root weight. Sindhu and Hans (1988) also reported that all wheat plants grown in pots containing *Eucalyptus tereticornis* leaf litter grew significantly less well than control plants. Bhaskar*et al* (1992) observed reduced results of seedling height and number of leaves of finger millet (*Eleusinecoracana*) due to powdered leaf litter of *Eucalyptus tereticornis*.

The utmost reduction in the mean root length of haricot bean was observed in the 200 g/pot of *Eucalyptus grandis* treated pots with value of 77.9%, while for maize it was observed in the 150 g/pot with value of 77.6%.At the same time the highest reduction in the men root length of haricot bean and maize with 76.7% and 76.4% respectively were observed in the 200 g/pot of *Eucalyptus camaldulensis* treated pots. Al-Juboory and Ahmad (1994) also reported that 2.5 and 5 kg leaf residues of *Eucalyptus camaldulensis* reduced weed growth especially *sorghum halepenses, cyperusrotundus* and *convolvulus arvensis*. Mizutani (1989) reported research results on the allelopathic effects of various plant extracts, which found that *Eucalyptus citrodora* and *Eucalyptus camaldulensis* inhibited growth of green foxtail, bamyard grass and rice. Srinivasan *et al* (1990) also observed the reduced crop germination and growth of *vignamungo, vignaradiata*, pigeonpea and soybean, when grown in the top soil taken from *Eucalyptus tereticornis*. Padhy*et al* (1992) tested the leaches of senescing and freshly fallen leaves of *Eucalyptus globulus* for their allelopathic effect on an improved cultivar of finger millet in laboratory. Germination, seedling shoots, and root growth were inhibited with the effects increasing with leachates concentration.

The maximum decreases in the mean collar diameter were observed in the 200 g/pot of *Eucalyptus grandis* treated pots with 85.2% and 76.9% for haricot bean and maize, and similarly in the 200 g/pot of *Eucalyptus camaldulensis* treated pots with values of 57.1% and 66.7% for haricot bean and maize respectively. These findings are confirmed by the work of (Ahmed*et al.*, 2008) who reported that leaf litters of *L. leucocephala* induced inhibitory effects on germination, growth and collar diameter of Falen (*Vignaunguiculata*), Chickpea (*Cicerarietinum*) and Arhor (*Cajanus cajan*). Anwar (1991) observed greater allelopathic effects of the fresh leaves extracts of *Eucaluptus alba, E. deglupta* and *E. robusta* on the growth of maize seedlings.

According to the mean result a 200 g/pot treatments of both *Eucalyptus* species reduced the highest in shoot, root fresh and shoot, root dry weight on haricot bean and maize crops as compared to other treatments. Findings by Lisanework and Michelson (1993) also reported that *Eucalyptus globules, Eucalyptus camaldulensis* and *Eucalyptus saligna* extract, reduces the fresh weight of maize seedlings. The results from (Khan *et al*, 1999) also showed that aqueous extract of *Eucalyptus camaldulensis* reduced seedlings dry weight, fresh shoot weight and fresh root weight of maize. Sunil and Khara (1991) also reported that water extract of leaves (green, brown and decayed) from 6 years old *Eucalyptus tereticornis* trees and bark were tested for their inhibitory effects on seed germination and primary root and shoot development of *phaseolus vulgaris* seedling.

12

CONCLUSION

The allelopathic compatibility of multipurpose tree species with companion agricultural crops may be crucial to determine the success of an agroforestry practice or system. *Eucalyptus* is claimed that it is notorious for having allelopathic effects on growth of agricultural crops growing in its vicinity. The allelopathic effect of *E. camaldulensis* and *E. grandis* were investigated on the soil reaction, seed emergence, and seedling growth of maize and haricot bean growing in the nursery. The leaf powder of both *Eucalyptus* species were prepared and applied to the pot soil in which both crops were grown. The effects on soil reaction, emergence, and growth of both crops were compared to a control with no leaf powder application. Results showed that the effect of leaf powder of both *Eucalyptus* species strongly reduced seed emergence and seedlings growth parameters of tested crops and decreased soil pH. It can be concluded from the results that allelochemicals present in the leaf powder of both *Eucalyptus* suppressed all the parameters measured in the crop species studied, and it was an increased inhibition with the increase of leaf powder application. Among the two tree species *Eucalyptus grandis* has more inhibitory effect on both crops as compared to *Eucalyptus camaldulensis*, and the negative effect of allelochemicals is more pronounced on haricot bean than maize in both tree species.

It is recommended that maize and haricot bean should not be planted very close to or in association with *Eucalyptus* trees due to the likely adverse effects (allelopathy) on seed emergence and other growth parameters. Thus, there is a need to provide information to farmers about where to plant or not when using the *Eucalyptus* species and their allelopathic effects on agricultural crops. Further studies are suggested to clarify the possible physiological mechanism related to allelopathic effects of the plants.

ACKNOWLEDGMENT

This study was financial supported by Rural Capacity Building Project (RCBP), Ministry of Agriculture, Ethiopia. I thank Dr. Abdu Abdulkadir and Dr. Tesfaye Abebe for their constructive suggestions and comments on the early version of this manuscript.

REFERENCES

Ahmed R, Hoque A. T. M R. and Hossain M K. 2008. Allelopathic effects of leaf litters of *Eucalyptus camaldulensis* on some forest and agricultural crops. Journal of Forestry Research 19: 19-24.

Al-juboory, B.A. and M. M. Ahmed. 1994. The allelopathic effects of some plant residues on some weed plants. Arab J. Plant Protection 12:3-10.

Anwar, C. 1991. Study of the allelopathic effect of Eucalyptus spp. On growth of maize seedlings. Bulletin-penelitian-Hutan. 547:9-17.

Bewley JD, M Black. 1994. Seeds: physiology of development and germination. Plenum Press, New York, 445 pp.

Bhaskar, V., A. Arali, and B.C. Shankara. 1992. Alleviation of allelopathic effects of Eucalyptus hybrid Through litter burning. In Proceedings Ins. Nat. Symp., Allelopathy in Agro-ecosystems, Feb 12-14, 1992.

CNST (Council of the National State of Tigray), 1997. The National State of Tigray Rural Land Proclamation No. 23/1997. Council of the National State of Tigray, Mekele.

Davidson, J. 1995. Eucalypts Tree Improvement and Breeding. Ministry of Natural Resources Development and Environmental Protection, Addis Ababa.

Ebrahim, E.E., H.A. Mohammad and A.F. Mustafa. 1999. Allelopathy effect of *Eucalyptus* and *Conocarpus*plantation on germination and growth of two sorghum species. Sudan J. Agric. Res. 2:9-14.

El-Darier. 2002. Allelopathic Effects of *Eucalyptus rostrata* on Growth, Nutrient Uptake and Metabolite Accumulation of *Viciafaba* L. and*Zea mays* L. Pakistan J. Journal of Biological Sciences. 5: 6-11.

El-Khawas, A.S. and M.M. Shehata. 2005. The allelopathic potentialities of *Acacia nilotica* and *Eucalyptus rostrata* on monocot (*Zea mays* L.) and Dicot (*Phaseolus vulgaris* L.) plants. Biotechnology, 4: 23-34.

Evans, J. 1992. Plantation Forestry in the Tropics: 2nd Edition. Oxford University Press, New York.

FAO. 1981. Eucalypts for Planting. FAO Forestry and Forest Products Studies 11. FAO, Rome.

Jagger, P. and Pender, J. 2000. The Role of Trees for Sustainable Management of Less Favored lands: The Case of Eucalypts in Ethiopia. International Food Research Institute, Washington.

Khan, M.A., I. Hussain and E.A. Khan. 2007. Effect of aqueous extract of *Eucalyptus camaldulensis*L. on germination and growth of maize. Pak. J. Weed Sci. Res. 13(3-4): 177-182.

Khan, R.A. 2003. Studies on weed control in sugarcane and allelopathic effects of *Eucalyptus* on field crops. Ph.D. Thesis, Gomal University, NWFP, Pakistan.

Kumar M, Singshi S, and Singh B. 2008. Screening indigenous tree species for suitable tree crop combinations in the agroforestry system of Mizoram, India. Estonian Journal of Ecology, 57, 4, 269-278.

Lisanework Nigatu, A Michelsen. 1993. Allelopathy in agroforestry systems the effects of leaf extracts of *Cupressuslusitanica* and three *Eucalyptus* spp. on four Ethiopian crops. Agroforestry Systems21: 63-74.

Mizutani, J. 1989. Plant allelochemicals and their rates. Forestry Abst. 1993. 7-G. Seed Abst.

Mkula, N.P. 2006. Allelopathic interference of silverleaf nightshade (*Solanumelaeagnifolium* Cav.) with the early growth of cotton (*Gossypiumhirsutum* L.). University of Pretoria, Pretoria.

Oyun M.B. 2006. Allelopathic Potentialities of *Gliricidiasepium*and *Acacia auriculiformis*on the Germination and Seedling Vigour of Maize (*Zea mays L.*). Akure, Nigeria , American Journal of Agricultural and Biological Science 1 (3): 44-47.

Padhy. B., P.A. Khan, B. Achariya and N. P. Bupripata. 1992. Allelopathic effects of eucalyptus leaves on Seed germination and seedling growth of finger millet. Proceedings Ind. Soc. Of Allelopathy, 102-104. Forestry Abst. 53 (10):7556, 1992.

Rao, N.S. and P.C. Reddy. 1984. Studies on the inhibitory effects of Eucalyptus (hybrid) leaf extracts on the germination of certain food crops. Indian Forester 110(2): 218-222.

Rho BJ, BS Kil. 1986. Influence of phytotoxin from *Pinusrigida* on the selected plants. J. Nat. Sci. 5: 19-27.

Saxena A, DV Singh, NL Joshi. 1996. Autotoxic effects of pearl millet aqueous extracts on seed germination and seedling growth. Journal of Arid Environments 33: 255-260.

Schumann, A.W., K. M. Little and N. S. Eccles. 1995. Suppression of seed germination and early seedling growth by plantation harvest residues. South African J. Plant and Soil 12: 170 – 172.

Shivanna, L.R., K. T. Prasanna, and J. Mumtaz. 1992. Allelopathic effect of eucalyptus, an assessment on the response of agricultural crops. Proceedings. 1st National Symposium. Allelopathy in Agroecosystem, Feb 12-14 1992.

Sindhu, D.S. and A. W. Hans. 1988. Preliminary studies on the effect of eucalyptus leaf litter on accumulation of biomass in wheat. J. Tropic. Forestry, 4(4):328-330. (Forestry Abst. 51 (21):7690., 1990).

Srinivasan, K., M. Ramswamy, and R. Shantha. 1990. Tolerance of pulse crops to allelochemicals of tree species. Indian J. Pulses Res., 3 (1): 40-44.

Sunil, P. and A. Khara. 1991. Allelopathic effect of *Eucalyptus terelicornis* on *Phaseolus vulgaris* seedling. International Tree Crops J. 6(4): 287-293.

Swaminathan, C., K. Sivagananam, and P. Srimathi. 1993. Allelopathic proclivities of multipurpose trees. My Forest 29:147-149.

Thakur, V.C. and S.D.Bhardwaj. 1992. Allelopathic effect of tree leaf extracts on germination of wheat and maize. Seed Research 20:153-154.